50 FORMULAS

THAT CHANGED THE WORLD

BALUNGI FRANCIS

Copyright © Balungi Francis, 2019

The moral right of the author has been asserted.

All rights reserved. Apart from any fair dealing for the purposes of research or private study or critism or review, no part of this publication may be reproduced, distributed, or transmitted in any form or by any means, including photocopying, recording, or other electronic or mechanical methods, or by any information storage and retrieval system without the prior written permission of the publisher.

Contents

Introduction 1
1 Heisenberg's uncertainty principle 5
2 Dirac equation 7
3 The Schrödinger equation 8
4 Bernhard Riemann's formula 10
5 Maxwell's laws 11
6 Second law of thermodynamics 13
7 Pi 15
8 The Euler-Lagrange equation 16
9 Newton's law of gravity 18
10 Einstein Energy –Mass equation 20
11 Einstein Field equation 22
12 Wave equation 24
13 The Yang-Baxter equation 25
14 Bayes' theorem 27
15 Planck – Einstein relation 29
16 The Pythagorean Theorem 30
17 Boltzmann entropy 32
18 Newton's second law 34
19 Euler's Identity 36
20 Heron's formula 38
21 The logistic map 40
22 Quaternion formula 41

23 The Black-Scholes model- The Midas 43
24 Shannon's information theory 45
25 The Navier –Stokes equation 46
26 Logarithms 48
27 Hodgkin-Huxley model 50
28 The Fourier transform 51
29 Calculus 52
30 The square root of minus one 53
31 The normal distribution 54
32 Euler's formula for polyhedra 55
33 The Google Formula/PageRank 56
34 Bernoulli's principle 58
35 The Butterfly effect 60
36 The Facebook Formula -EdgeRank 61
37 The law of light reflection 63
38 Snell-Descartes law 64
39 Hooke's law 65
40 Ideal gas law 67
41 The standard model 68
42 The first Friedmann equation 73
43 The Callan-Symanzik equation 74
44 The minimal surface equation 76
45 The Heat equation 77
56 The Gaussian Integral 78
47 The Basel Identity 79

48 The KFC formula	80
49 The Coca-Cola formula	83
50 The Kalman Filter	85
My Favorite Formula	87
Bibliography/References	88

Introduction

Have you ever looked up into the night sky and wondered just how many stars there are in space? This question has fascinated scientists as well as philosophers, musicians and dreamers throughout the ages. Have you tasted the Kentucky Fried Chicken lately, and marveled at the secret formula behind the finger licking good chicken? What is the formula behind the Coca-Cola drink, Face book, Google, and the theory of everything, I can't stop to look but the list is endless, this book presents these and many other formulas which changed and still run the world.

So how many stars are there in the universe? It is easy to ask this question, but difficult for scientists to give an answer. But the beauty of equations and formulas is twofold. Just as there exist a formula for KFC's delicious "finger lickin' good chicken, there is also a formula hidden deep in the cosmos that awaits one or two minds to grasp, and surely this equation will be used to count the number of stars in our mother universe. However we are not far away from the truth, the ESA's infrared space observatory Herschel have devised a formula to count the stars in the universe to about 10^{24} stars not to mention but the Drake equation is about to estimate the number of active, communicative extraterrestrial civilizations in the Milky way galaxy and the Universe at large. These equations will someday prove whether we are alone in the universe and the power of formulas will be revealed in a different fashion.

Equations and formulas are the lifeblood of mathematics, physics, science and technology. While a formula is a mathematical relationship or rule expressed in symbols or a list of ingredients with which something is made, an equation is rather a statement that the values of two mathematical expressions are equal (or the process of equating one thing with another). Without equations or formulas our world would not exist in its present form.

Whether it's the construction of your house, the layout of streets in your neighborhood, the simple act of starting your car or turning on your dishwasher, when you play the piano or searching the internet, complex and basic formulas are everywhere. I want to convince you that equations have played a vital part
In creating today's world, from mapmaking to satnav, from music to television, from discovering America to exploring the moons of Jupiter, from telegram to instagram and smartphones.

Equations are an essential tool for describing how many things in the natural world function and interact. But some equations have had a more far-reaching influence than others. Below I present 50 such equations that changed and are still shaping the world.

Complex equations with many unknowns, radical mathematical theorems dating back to antiquity, to late twentieth century discoveries, has all shaped our world. And with each new concept, our understanding of the physical world around us grows.

People understood geometry and algebra by about 2000 BCE. Around this time, both the Babylonians and Ancient Egyptians were aware of the number (pi) the ratio of a circle's circumference to its diameter. By about 1500 BCE, the Babylonians were also aware of Pythagoras' theorem, which shows how the lengths of the sides of right-angled triangles are related.

The theorem is named after ancient Greek mathematician Pythagoras (though some say the concept predates him) which shows that, although all math theorems and formulae simply exist and are waiting to be discovered, we can at least praise some individuals for finding them or working them out.

Kepler was also inspired by Pythagoras, and believed that the motion of the planets produces music. He used mathematics to show that the planets orbit the Sun in ellipses and, by 1619, he was able to determine the time it takes each planet to orbit and their relative distances from the Sun.

In 1687, Newton published his law of universal gravitation. This was groundbreaking because it showed, not just that abstract mathematical principles, such as the newly invented calculus, could be applied to what we observe in nature, but that the laws responsible for the movement of the planets are also responsible for the movement of objects on Earth. Newton also believed that the universe could be understood as a mathematical object, and described God as "skilled in mechanics and geometry".

Newton's contemporary, Leibniz, discovered another link between mathematics and nature when he first considered the idea of fractals. Twentieth-century mathematicians, such as French mathematician Gaston Julia and Polish-French-American mathematician Benoit Mandelbrot, were inspired by Leibniz to create complicated fractals of their own.
By this time, quantum mechanics, and German-Swiss-American physicist Albert Einstein's theories of special and general relativity, had shown that nature obeys the laws of mathematics, even when this contradicts our common sense understanding of the world.

So, as we can see, Maths and physics has been ever-present throughout the history of Man, yet a number of significant breakthroughs have emerged thanks to skilled mathematicians who have come to find them.

We will look at some of the most famous formulas and equations in maths, physics, economics, technology, the internet, computer science, catering (cookery, food preparation) and beverage.

If you wonder why formulas are so important, and the impact that each major formula or equation has brought, read on to discover 50 revolutionary formulas to take your maths, physics, economics, computer science applications, catering and beverage intuition to another level. There is, of course, no end to the number of mathematical formulas and expressions that exist (some might say the list is endless), but in this book we focus on some of the better known algebraic equation and formulae and provide some helpful notation.

1 Heisenberg's uncertainty principle

$$\Delta P \Delta x \geq \frac{\hbar}{2}$$

What does it say?

It states that the more certain you are about a particle's momentum (P) the less certain you are about the particle's position (x) that is, momentum and position can never both be known exactly.

What did it teach us?

One can never know with perfect accuracy both of those two important factors which determine the movement of one of the smallest particles, that is its position and its velocity. It is impossible to determine accurately both position, direction and speed of a particle at the same instant.

But was it practical?

It is used in testing number-phase uncertainty relations in superconducting and or quantum optics systems. It is also applied to low–noise technology such as that required in gravitational wave interferometers. It also opened a door to wave mechanics and the Schrödinger equation.

History

The uncertainty principle was established by Werner Heisenberg in 1927 in his paper "on the perceptual content of quantum theoretical kinematics and mechanics", Heisenberg established this expression as the minimum amount of unavoidable momentum disturbance caused by any position in measurement.

2 Dirac equation

$$(i\partial - m)\psi = 0$$

What does it say?

The Dirac equation unifies quantum mechanics
(Which describes the behavior of tiny objects) with Einstein special theory of relativity (which describes the behavior of fast-moving objects).

What did it teach us?

Describes how particles like electrons behave when they travel close to the speed of light. It was the first step towards what's called quantum field theory.

But was it practical?

The Dirac equation predicted the existence of antimatter-the mirror image of all known particles.

History

The equation was derived by Paul Dirac in 1928.

3 The Schrödinger equation

$$H(t)|\psi(t)\rangle = i\hbar \frac{\partial}{\partial t}|\psi(t)\rangle$$

What does it say?

As Einstein's theory of general relativity helped explain the universe on a large scale, this equation sheds light on the behaviour of atoms and subatomic particles.

What did it teach us?

The Schrödinger equation explains the changes over time of a particle. It describes the states of a particle, from which it is possible to describe any state. This equation poses a real philosophical question: Is matter made up of the presence of possible physical states (solids, liquids, gases)?

But was it practical?

The application of this equation can be found in modern technology including nuclear energy, solid-state computers and lasers. It also helps keep cats in a form of suspended animation between life and death, apparently. But seriously, this equation literally changed the field of quantum physics forever. it is a linear partial differential equation that describes the wave function of a quantum-mechanical system. its discovery was a significant landmark in the development of quantum mechanics

History

The equation is named after Erwin Schrodinger, who postulated the equation in 1925, and published it in 1926, forming the basis for the work that resulted in his Nobel Prize in physics in 1933.

4 Bernhard Riemann's formula

$$f(x) = li(x) - \sum_{\rho} li(x^\rho) - log(2)$$
$$+ \int_x^\infty \frac{dt}{t(t^2-1)log(t)}$$

What does it say?

Riemann's equation reveals that prime numbers are controlled by something called the zeta function. It allows you to calculate the number of primes below a given number.

What did it teach us?

This equation implies that there is some deeper rules governing which numbers are prime. For instance, the Riemann equation reveals that there are 25 primes between 1 and 100.

But was it practical?

Prime numbers are of huge practical importance, because most cryptography relies on them. "Today all the codes used on the internet exploit the primes to keep messages secure, Unlock the secret of the primes and it's possible all these codes too will be unlocked."

History

The mathematician Bernhard Riemann published this equation in 1859.

5 Maxwell's laws

$$\nabla \cdot D = \rho$$
$$\nabla \cdot B = 0$$
$$\nabla \times E = -\frac{\partial B}{\partial t}$$
$$\nabla \times H = J + \frac{\partial D}{\partial t}$$

What do they say?

Maxwell's equations describe how electric charges interact, as well as explaining electric currents and magnetic fields. They say that a changing magnetic field produces an electric field and vice versa.

What did it teach us?

Maxwell's equations, also called Maxwell-Lorentz equations, are fundamental laws of physics. They underpin our understanding of the relationship between electricity and magnetism, and are among the essential, fundamental laws of modern physics. They show that electricity and magnetism are intimately interrelated.

But was it practical?

Because Maxwell's equations predict that electromagnetic waves exist and move at a constant speed of light, therefore they predict that light is an electromagnetic wave. They lead to the invention of radio, radar, television, wireless connections for computer equipment, and most modern communications.

History

The equations are named after that physicist and mathematician James Clerk Maxwell, who published an early form of the equations that included the Lorentz force law between 1861 and 1862. Maxwell first used the equations to propose that light is an electromagnetic phenomena.

6 Second law of thermodynamics

$$dS \geq 0$$

What does it say?

This states that, in a closed system, entropy (S) is always steady or increasing. Thermodynamic entropy is, roughly speaking, a measure of how disordered a system is. A system that starts out in an ordered, uneven state say, a hot region next to a cold region will always tend to even out, with heat flowing from the hot area to the cold area until evenly distributed.

What did it teach us?

It helps us, among other things, to understand the direction of heat transfer. This theory can be expressed in terms of the change in entropy of a system (dS). In this equation, dS is calculated by measuring how much heat has entered a closed system (δQ) divided by the common temperature (T) at the point where the heat transfer took place.

But was it practical?

Better steam engines, estimates of the efficiency of renewable energy, the heat death of the universe, proof that matter is made of atoms, and paradoxical connections with the arrow of time.

History

Rudolf Clausius was the first person to formulate the second law in 1850 after recognizing the significance of James Prescott Joule's work on the conservation of energy of 1824. He then provided the definition of the law in 1865.

7 Pi

$$\pi = \frac{C}{d}$$

What does it say?

It simply describes how the circumference of a circle varies with its diameter. The ratio of the two is a number called pi. It is roughly 3.14159, but not exactly: pi is an irrational number, meaning the digits go on forever without repeating.

What did it teach us?

Its importance is in the calculation of the radius of the circle, the area of a circle, the surface area of a sphere and its volume. Pi became useful in ancient civilations during the construction of prymadis and other structures which required an accurate determination of Pi.

But was it practical?

Pi is incredibly important number.We have to calculate it to very high precision for modern technology such as GPS to work at all. It can be used to describe the geometry of the world. It is also applied in the production of CDs.

8 The Euler-Lagrange equation

$$L_x(t, q(t), q'(t)) - \frac{d}{dt} L_v(t, q(t), q'(t)) = 0$$

What does it say?

It's a recipe to generate an infinite variety of possible physical laws

What did it teach us?

The way that all classical physics can be expressed and understood within a single framework like that helps reveal deep links between seemingly different phenomena.

But was it practical?

This equation is used to analyse everything from the shape of a soap bubble to the trajectory of a rocket around a black hole.

History

It was developed by Swiss mathematician Leonhard Euler and Italian mathematician Joseph-Louis Lagrange in the

1750s. This came as a result of the study of the tautochrone problem. This is the problem of determining a curve on which a weighted particle will fall to a fixed point in a fixed amount of time, independent of the starting point.

9 Newton's law of gravity

$$F = G\frac{Mm}{R^2}$$

What does it say?

Newton's law of universal gravitation states that every particle attracts every other particle in the universe with a force which is directly proportional to the product of their masses and inversely proportional to the square of the distance between their centers.

What did it teach us?

The force of gravity governs the movement of bodies in the solar system. It is this simple mathematical law which determines the motion of bodies.

But was it practical?

The force of gravity accurately predicts the planetary orbits, it was used to put the first man on the moon, it predicts the return of comets, the rotation of galaxies, the solar eclipses, artificial satellites, satellite communications and television, the GPS and interplanetary probes. I almost forgot, it is why NASA was established in the first place.

History

It was published by Isaac Newton in his principia on 5th July 1687. In 1986 Robert Hooke reported to the Royal society that Newton had stolen his inverse square law. The war between Hooke and Newton has never been resolved up to this day.

10 Einstein Energy –Mass equation

$$E = mc^2$$

What does it say?

For particles moving at a constant speed of light, the mass of a particle can be converted to energy and vice versa is true.

What did it teach us?

A small amount of mass can be converted into an enormous amount of energy such as those in nuclear weapons- atomic bombs

But was it practical?

Einstein's famous equations on relativity not only answered many previously unsolved questions, but it also helped change the way we look at time, space, and gravity. It is used to assist in explaining everything from black holes to the big bang, to nuclear power, atomic bombs (like the one that was dropped on Hiroshima and Nagasaki in Japan).

History

It was formulated by Einstein on 21st November 1905 in his paper "Does the inertia of a body depend upon its energy content". While he was the first to have correctly deduced the mass energy equivalence formula, he was not the first to have related energy with mass. Albert Einstein did not formulate exactly the formula given above rather he uses L as the energy in the form of radiation in his 1905 paper.

11 Einstein Field equation

$$R_{\mu\nu} - \frac{1}{2} R g_{\mu\nu} + \Lambda g_{\mu\nu} = k T_{\mu\nu}$$

What does it say?

The equation relates the geometry of space-time with the distribution of matter within it. The left hand side of the equation represents the geometry or curvature of space-time while the right hand side represents the matter-energy density. It is the main result of Einstein General relativity theory (1915-1916).

What did it teach us?

Matter in the space-time causes space to curve and in return space-time tells matter how to move. The motion of bodies in space is nolonger described by the Newton's force of gravity but rather with the curved nature of space-time. Space and time are not separate entities but are united in such a way that time is a fourth dimension.

But was it practical?

The solution to the equation predict gravitational time dilation, gravitational lensing, gravitational redshift of light, gravitational time delay, existence of gravitational waves, expanding universe, existence of black holes and in GPS applications.

12 Wave equation

$$\frac{\partial^2 u}{\partial t^2} = c^2 \nabla^2 u$$

What does it say?

The wave equation is a second order partial differentiation equation that describes the propagation of waves. It relates the change of propagation of the wave in time to the change of propagation in space and a factor of the wave speed (v) squared.

What did it teach us?

It applies to all kinds of waves, from water waves to sound and vibrations, and even light and radio waves.

But was it practical?

It began with a simple model of a vibrating violin string, and developed into something used to study a diverse range of phenomena, from earthquakes to oil prospecting and even safety on ships. Its link to music helped explain how our ears hear sound, and why some combinations are harmonious while others are jarring.

13 The Yang-Baxter equation

$$(\mathbb{R} \otimes \mathbb{I})(\mathbb{I} \otimes \mathbb{R})(\mathbb{R} \otimes \mathbb{I}) = (\mathbb{I} \otimes \mathbb{R})(\mathbb{R} \otimes \mathbb{I})(\mathbb{I} \otimes \mathbb{R})$$

What does it say?

Also called the star-triangle relation is a consistency equation which was first introduced in the field of statistical mechanics. It states that a matrix $\mathbb{R}$, acting on two out of three objects, satisfies the Baxter equation.

What did it teach us?

It depends on the idea that in some scattering situations, particles may preserve their momentum while changing their quantum internal states. The Yang-Baxter equation also shows up when discussing knot theory and braid groups where $\mathbb{R}$ corresponds to swapping two strands.

But was it practical?

Like the Euler-Lagrange equation, it looks simple but has profound implications for many areas of mathematics and physics. These include how waves behave in shallow water, the interaction of subatomic particles, the mathematical theory of knots, and string theory.

History

It takes its name from independent work of C.N.Yang from 1986, and R.J. Baxter from 1971.

14 Bayes' theorem

$$P(A/B) = \frac{P(B/A).P(A)}{P(B)}$$

What does it say?

It calculates the probability that one event (A) is true, given that another event (B) is also true.

What did it teach us?

In probability theory and statistics , the equation describes the probability of an event, based on prior knowledge of conditions that might be related to the event. For example if AIDs is related to gender, then, using Bayes theorem , a person's gender can be used to more accurately assess the probability that they have AIDs than can be done without knowledge of the person's gender.

But was it practical?

It is used for many purposes, including detecting faults, surveillance, military defence, search-and-rescue operations, medical screening and even spam filters.

History

It is named after Thomas Bayes(1701-1961). Bayes never published his manuscript which contained the formula. The manuscript was later edited by Richard Price and presented to the royal society and later appeared in philosophical transactions in 1763.

15 Planck – Einstein relation

$$E = hf$$

What does it say?

It states that the energy of a photon (E) is proportional to its frequency (f).

What did it teach us?

This law accounts for the quantized nature of light. This law states that electromagnetic energy can only be emitted or absorbed in specific (quantised) amounts. This is now known to be due to electromagnetic radiation not being a continuous wave but actually many photons, "packets of light".

But was it practical?

This simple law was used to understand phenomena such as the photoelectric effect, black-body radiation, Compton Effect, deBrogile wavelength and this equation is now remembered as the birth of quantum theory.

16 The Pythagorean Theorem

$$a^2 + b^2 = c^2$$

What does it say?

This is surely one of the best-known theorems. The Pythagorean theorem relates the sides of a right triangle, where a and b are the lengths of the legs and c is the length of the hypotenuse. It also relates triangles to squares.

What did it teach us?

This equation is essential to an understanding of geometry and trigonometry, and indeed has shaped our understanding of those branches of mathematics. Thanks to Pythagoras and his famous equation, it's now easy to calculate lengths, angles and to demonstrate that a given triangle is right-angled.

But was it practical?

This equation is one that actually changed the world. It enabled us to be able to formulate better maps and help find the shortest distance between things; amongst other things. It is also used heavily in architecture, woodworking, and many other fields such as, surveying, navigation, general relativity theory etc

17 Boltzmann entropy

$$S = k_B \ln W$$

What does it say?

It is a key equation for statistical mechanics formulated by Ludwig Boltzmann. It relates the entropy of a macrostate (S) to the number of microstates corresponding to that macrostate (W).

What did it teach us?

The key thing here is that multiple different microstates can correspond to the same macrostate. Therefore, a simpler statement would be that the entropy is related to the arrangement of particles within the system (or the 'probability of the macrostate').

But was it practical?

The equation shows the relationship between entropy and the number of ways the atoms or molecules of a thermodynamic system can be arranged. This equation can be used to derive thermodynamic equations such as the ideal gas law

History

The equation was originally formulated by Ludwig Boltzmann between 1872 and 1875, but later put into its current form by Max Planck in about 1900.

18 Newton's second law

$$F = ma$$

What does it say?

The vector sum of the forces (F) on an object is equal to the mass (m) of that object multiplied by the acceleration (a) of the object.

What did it teach us?

It is the cornerstone of classical mechanics, which allows the motion of objects subjected to forces to be calculated.

But was it practical?

It has been applied to a huge number of problems from the motion of cars all the way up to the orbits of the planets around our sun. It was only superseded by the theory of quantum mechanics in the early 1900s.

History

The second law appears in Newton's "pricipia" which was first published in 1687.

19 Euler's Identity

$$e^{i\pi} + 1 = 0$$

What does it say?

Euler's identity is a very famous equation which relates the seemingly random values of π, e, and the square root of -1. It is considered by many to be the most beautiful equation in mathematics.

What did it teach us?

It makes use of three fundamental operations in arithmetic: addition, multiplication and exponentiation.

But was it practical?

This equation decorates the Palais de la Decouverte in Paris, it paved the way for the development of topology, a branch of modern math. The numbers in the Euler identity all have many practical applications, including communication, navigation, energy, manufacturing, finance, meteorology and medicine.

History

The equation was published by Leonhard Euler in 1755. It applied in the case of a perfect fluid. Euler was the most prolific mathematician of all time and the Mozart of mathematics. Most of modern mathematics and physics derives from work of Leonhard Euler

20 Heron's formula

$$A = \sqrt{s(s-a)(s-b)(s-c)}$$

$$S = \frac{a+b+c}{2}$$

What does it say?

Let ΔABC be the triangle with sides a = 4, b = 13 and c = 15. The semi perimeter is $s = \frac{1}{2}(4 + 13 + 15) = 16$, and the area is $\sqrt{576} = 24$.

What did it teach us?

Unlike other triangle area formulae where one needs to calculate angles or other distances in the triangle first. The Heron formula gives the area of a triangle when the lengths of all three sides are known.

But was it practical?

It is used heavily in architecture, land surveying, woodworking, and many other fields such as, navigation, general relativity theory etc

History

The formula is credited to Heron of Alexandria, and a proof can be found in his book, *Metrica*, written circa. CE 60. It has been suggested that Archimedes knew the formula over two centuries earlier, and since *Metrica* is a collection of the mathematical knowledge available in the ancient world, it is possible that the formula predates the reference given in that work.

21 The logistic map

$$x_{n+1} = rx_n(1 - x_n)$$

What does it say?

The equation estimates the change in a population of creatures across generations with limited resources.

What did it teach us?

It helped in the development of chaos theory, which has completely changed our understanding of the way that natural systems work.

But was it practical?

It models earthquakes and weather forecast.

History

Robert May was the first to point out that this model of population growth could produce chaos in 1975. Important work by mathematicians Vladimir Arnold and Stephen Smale helped with the realization that chaos is a consequence of differential equations.

22 Quaternion formula

$$i^2 = j^2 = k^2 = ijk = -1$$

What does it say?

These are number systems that extend the complex numbers. Hamilton defined them as the quotient of two directed lines in a three-dimensional space or as the quotient of two vectors.

What did it teach us?

This equation describes how to work with complex numbers that include the square roots of negative numbers. Nowadays, quaternion algebra is central to the computer graphics industry. It is used to describe the orientations of objects on the screen.

But was it practical?

Quaternions have practical uses in three-dimensional computer graphics, computer vision, crystallographic texture analysis, Euler angles and rotation matrices.

History

The formula is credited to William Rowan Hamilton who first described it in 1843. It is believed that on his way to the Royal Irish Academy, the concepts of quaternions came to him and he was forced to carve the formula into the stone of Brougham Bridge as he paused on it.

23 The Black-Scholes model- The Midas

$$\frac{1}{2}\sigma^2 S^2 \frac{\partial^2 V}{\partial S^2} + rS\frac{\partial V}{\partial S} + \frac{\partial V}{\partial t} - rV = 0$$

What does it say?

It prices a derivative based on the assumption that it is riskless and that there is no arbitrage opportunity when it is priced correctly.

What did it teach us?

It helped create the now multi trillion dollar derivatives market. It is argued that improper use of the formula contributed to the financial crisis. In particular, the equation maintains several assumptions that do not hold true in real financial markets.

But was it practical?

Variants are still used to price most derivatives, even after the financial crisis.

History

The equation was developed by Fischer Black and Myron Scholes, then expanded by Robert Merton. The latter two won the 1997 Noble Prize in Economics for the discovery.

24 Shannon's information theory

$$H = -\sum p(x) \log p(x)$$

What does it say?

The information theory estimates the amount of data in a piece of code by probabilities of its component symbols.

What did it teach us?

This equation established the boundaries that made everything from CDs to digital communication possible.

But is it practical?

It is an efficient error-detecting and error-correcting codes, used in everything from CDs to space probes. Applications include statistics, artificial intelligence, cryptography, and extracting meaning from DNA sequences.

History

It was developed by Bell Labs engineer Claude Shannon in the years after World war2.

25 The Navier –Stokes equation

$$\rho\left(\frac{\partial v}{\partial t} + v.\nabla v\right) = -\nabla p + \nabla.T + f$$

$$yt + A(y)y_x = 0$$

What does it say?

The left hand side of the equation is the acceleration of a small amount of fluid, the right hand side indicates the forces that act upon it. It represents the flow of incompressible fluid

What did it teach us?

Once computers became powerful enough to solve this equation, it opened up a complex and very useful field of physics. It is particularly useful in making vehicles more aerodynamic.

But is it practical?

The equation can be used to model things such as weather, ocean currents, and flow of hot air. It allowed for the development of modern passenger jets.

History

Leonhard Euler made the first attempt at modeling fluid movement, French engineer Claude –Louis Navier and Irish mathematician George Stokes made the leap to the model still used today. Even though scientists have successfully used the equations in real life applications, no one has ever been able to prove its correctness. There's an outstanding $1 million bounty for the prover of the equations.

26 Logarithms

$$\log xy = \log x + \log y$$

What does it say?

It means that one can multiply numbers by adding related numbers.

What did it teach us?

Logarithms were revolutionary in making calculation faster and more accurate for engineers and astronomers. That's less important with the advent of computers, but they're still an essential to scientists. It taught us that addition is much simpler than multiplication.

But was it practical?

It was an efficient method for calculating astronomical phenomena such as eclipses and planetary orbits, quick method for scientific calculations, radioactive decay and psychophysics of human perception.

History

The initial concept was discovered by the Scottish Laird John Napier of Merchiston in an effort to make the multiplication of large numbers, then incredibly tedious and time consuming, easier and faster. It was later refined by Henry Briggs to make reference tables easier to calculate and more useful.

27 Hodgkin-Huxley model

$$I = C_m \frac{dV_m}{dt} + g_K(V_m - V_K) + gN_a(V_m - V_{Na}) + g_l(V_m - V_l)$$

What does it say?

It is a mathematical model that describes how action potentials in neurons are initiated and propagated. It is a set of nonlinear differential equations that approximates the electrical characteristics of excitable cells such as neurons and cardiac myocytes.

What did it teach us?

It models the way nerve cells send signals to each other.

But is it practical?

It formed the basis of theoretical neuroscience.

History

The model was described by Alan Hodgkin and Andrew Huxley in 1952 to explain the ionic mechanism underlying the initiation and propagation of action potentials in the squid giant axon. They received the 1963 Nobel Prize in Physiology or Medicine for their discovery.

28 The Fourier transform

$$\hat{f}(\xi) = \int_{-\infty}^{\infty} f(x) e^{-2\pi i x \xi} \, dx$$

What does it say?

It describes patterns in time as a function of frequency.

What did it teach us?

The equation allows for complex patterns to be broken up, cleaned up, and analyzed .It is a gist for many signal analysis.

But is it practical?

It is used in DNA, to compress information for the JPEG image format,to clean up old or damaged audio recordings, to analyze earthquakes. Today it is used to store fingerprint data efficiently and to improve medical scanners.

History

This equation was discovered by Joseph Fourier as an extension from his famous heat flow equation, and the previously described wave equation.

29 Calculus

$$\frac{df}{dt} = \lim_{h \to 0} \frac{f(t+h) - f(t)}{h}$$

What does it say?

This simple equation allowed for the calculation of an instantaneous rate of change.

What did it teach us?

Calculus is the source of differential equations, it is essential in our understanding of how to measure solid, curves, and areas.

But is it practical?

Applied in economics, computer science, weather forecast, surveying, medicine, calculation of tangents, areas, Newton's laws of motion, differential equations.

History

Calculus was discovered by Isaac Newton and Gottfried Leibniz although it is still not clear who discovered it first. It is believed that Newton plagiarized Leibniz work although no one knows the truth up to this day.

30 The square root of minus one

$$i^2 = -1$$

What does it say?

The square of an imaginary number is negative.

What did it teach us?

Imaginary numbers allow for complex analysis, which allows engineers to solve practical problems working in the plane.

But is it practical?

It is used in electrical engineering, complex mathematic theory, used to understand waves, heat, electricity, and magnetism.

History

Imaginary numbers were originally posited by famed gambler/ mathematician Girolamo Cardano, then expanded by Rafael Bombelli and John Wallis. They still existed as a peculiar, but essential problem in math until William Hamilton described this definition.

31 The normal distribution

$$\Phi(x) = \frac{1}{\sqrt{2\pi}\sigma} e^{\frac{(x-\mu)^2}{2\sigma^2}}$$

What does it say?

It defines the standard normal distribution, a bell shaped curve in which the probability of observing a point is greatest near the average, and declines rapidly as one moves away.

What did it teach us?

The equation is the foundation of modern statistics. Science and social science would not exist in their current form without it.

But was it practical?

It was used to determine whether drugs are sufficiently effective relative to negative side effects in clinical trials.

History

The initial work was by Blaise Pascal, but the distribution came into its own with Bernoulli. The bell curve as we currently know it, comes from Belgian mathematician Adolphe Quetelet.

32 Euler's formula for polyhedra

$$F - E + V = 2$$

What does it say?

It describes a space's shape or structure regardless of alignment.

What did it teach us?

It was fundamental in the development of topography, which extends geometry to any continuous surface. An essential tool for engineers and biologists.

But was it practical?

It was used to understand the behavior and function of DNA.

History

The relationship was first described by Descartes, then refined, proved, and published by Leonhard Euler in 1750.

33 The Google Formula/PageRank

$$R'(u) = c \sum_{v \in B_u} \frac{R'(v)}{N_v}$$

What does it say?

PageRank is a proprietary mathematical formula or algorithm that Google uses to calculate the importance of a particular web page/URL based on incoming links.

What did it teach us?

Through the pagerank formula, we are able to order search results so that more important and central web pages are given preference. In experiments, this turns out to provide higher quality search results to users. PageRank is therefore important because it will determine if your site shows up first or last when a potential customer looks for your keywords.

But was it practical?

It is used by Google by counting the number and quality of links to a page to determine a rough estimate of how important the website is. It is also used by facebook, Amazon, twiter, search engines and website development.

History:

Larry page and Sergey Brin developed PageRank at Stanford university in 1996 as part of a research project about a new kind of search engine.

34 Bernoulli's principle

$$P_1 + \frac{1}{2}\rho V_1^2 + \rho g h_1 = P_2 + \frac{1}{2}\rho V_2^2 + \rho g h_2$$

What does it say?

It states that within a horizontal flow of fluid, points of higher fluid speed will have less pressure than points of slower fluid speed.

What did it teach us?

If a fluid is flowing horizontally and along a section of a streamline, where the speed increases it can only be because the fluid on that section has moved from a region of higher pressure to a region of lower pressure; and if its speed decreases, it can only be because it has moved from a region of higher pressure.

But was it practical?

It applies to airfoil(aerodynamics), curve of a baseball, venture meter, an injector on a steam locomotive, the pitot tube and static port on an aircraft used to determine the airspeed of the aircraft, De Laval nozzle, carburetor, Bernoulli grip and in swinging of a cricket ball.

History

The principle is named after Daniel Bernoulli who published it in his book Hydrodynamica in 1738. The equation was derived by Leonhard Euler to its original form in 1752.

35 The Butterfly effect

$$|\delta Z(t)| \approx e^{\lambda t}|\delta Z_0|$$

What does it say?

The chaos theory describes how a small change in one state of a deterministic nonlinear system can result in large differences in a later state. As a metaphor, a butterfly flapping its wings in china can cause a hurricane in Texas.

What did it teach us?

According to Edward Lorenz, Chaos is when the present determines the future, but the approximate present does not approximately determine the future. This theory effectively helps us deal with complex systems whose behavior is highly sensitive to slight changes in conditions so that small alterations can give rise to unintended consequences. Chaotic behavior exists in many natural systems, including fluid flow, heartbeat irregularities, weather and climate.

But was it practical?

It is applicable to geology, mathematics, microbiology, biology, computer science, economics, engineering, finance, algorithmic trading, meteorology, population dynamics, robotics, and cryptography

36 The Facebook Formula - EdgeRank

$$\sum_{edges\ e} u_e\, w_e\, d_e$$

What does it say?

EdgeRank is a proprietary mathematical formula or algorithm that Facebook uses to determine what articles should be displayed in a user's News feed.

What did it teach us?

The algorithm hides boring stories, so if your story doesn't score well, no one will see it. Every action you make on facebook is called an edge. That means whenever a friend posts a status update, comments on another status update, tags a photo, joins a fan page, or RSVP's to an event it generates an "Edge", and a story about that Edge might show up in the user's personal newsfeed

But was it practical?

It is used by facebook to decide which stories appear in each user's newsfeed. It is also used in facebook advertising platforms where one is forced to promote his page in case it doesn't reach any significant audience.

History:

EdgeRank was developed and implemented by Serkan Piantino.

37 The law of light reflection

$$\theta_i = \theta_r$$

What does it say?

The angle at which the wave is incident on the surface equals the angle at which it is reflected.

What did it teach us?

In acoustics, reflection causes echoes and is used in sonar. In geology, it is important in the study of seismic waves.

But was it practical?

Reflection of VHF and higher frequencies is important for radio transmission and for radar. Even hard X-rays and gamma rays can be reflected at shallow angles with special "grazing" mirrors.

38 Snell-Descartes law

$$\boxed{\frac{\sin\theta_2}{\sin\theta_1} = \frac{v_2}{v_1} = \frac{n_1}{n_2}}$$

What does it say?

The ratio of the sines of the angles of incidence and refraction is equivalent to the ratio of phase velocities in the two media, or equivalent to the reciprocal of the ratio of the indices of refraction.

What did it teach us?

The law is used to determine the direction of light rays through refractive media with varying indices of refraction.

But was it practical?

The law is used in ray tracing to compute the angles of incidence, it is also used to find the refractive index of a material. The law is also satisfied in metamaterials, which allow light to bent backward at a negative angle of refraction with a negative refractive index.

39 Hooke's law

$$F_s = kx$$

What does it say?

The strain in a solid is proportional to the applied stress within the elastic limit of that solid.

What did it teach us?

Hookes law is important because it helps us understand how a stretchy object will behave when it is stretched or compacted.

But was it practical?

It is the foundation of seismology, molecular mechanics and acoustics. It is also the fundamental principle behind the spring scale, automotive suspension systems, hand sheers, wind-up toys, watches, rap traps, digital micromirror devices, the manometer, and the balance wheel of the mechanical clock.

History:

The law was stated by Robert Hooke in 1676 as a Latin anagram. He published the solution of his anagram in 1678.

40 Ideal gas law

$$PV = nRT$$

What does it say?

The state of an amount of gas is determined by its pressure, volume, and temperature.

What did it teach us?

It is a good approximation of the behavior of many gases under many conditions.

But was it practical?

It is used in stoichiometry, refrigerator and air conditioners.

History:

It was first stated by Emile Clapeyron in 1834 as a combination of the empirical Boyle's law, Charles's law, Avogadro's law and Gay-Lussac's law.

41 The standard model

$$-\tfrac{1}{2}\partial_\nu g_\mu^a \partial_\nu g_\mu^a - g_s f^{abc}\partial_\mu g_\nu^a g_\mu^b g_\nu^c - \tfrac{1}{4}g_s^2 f^{abc}f^{ade}g_\mu^b g_\nu^c g_\mu^d g_\nu^e +$$
$$\tfrac{1}{2}ig_s^2(\bar{q}_i^\sigma \gamma^\mu q_j^\sigma)g_\mu^a + \bar{G}^a \partial^2 G^a + g_s f^{abc}\partial_\mu \bar{G}^a G^b g_\mu^c - \partial_\nu W_\mu^+ \partial_\nu W_\mu^- -$$
$$M^2 W_\mu^+ W_\mu^- - \tfrac{1}{2}\partial_\nu Z_\mu^0 \partial_\nu Z_\mu^0 - \tfrac{1}{2c_w^2}M^2 Z_\mu^0 Z_\mu^0 - \tfrac{1}{2}\partial_\mu A_\nu \partial_\mu A_\nu - \tfrac{1}{2}\partial_\mu H \partial_\mu H -$$
$$\tfrac{1}{2}m_h^2 H^2 - \partial_\mu \phi^+ \partial_\mu \phi^- - M^2\phi^+\phi^- - \tfrac{1}{2}\partial_\mu \phi^0 \partial_\mu \phi^0 - \tfrac{1}{2c_w^2}M\phi^0\phi^0 - \beta_h\left[\tfrac{2M^2}{g^2}+\right.$$
$$\tfrac{2M}{g}H + \tfrac{1}{2}(H^2+\phi^0\phi^0+2\phi^+\phi^-)\Big] + \tfrac{2M^4}{g^2}\alpha_h - igc_w[\partial_\nu Z_\mu^0(W_\mu^+ W_\nu^- -$$
$$W_\nu^+ W_\mu^-) - Z_\nu^0(W_\mu^+ \partial_\nu W_\mu^- - W_\mu^- \partial_\nu W_\mu^+) + Z_\mu^0(W_\nu^+ \partial_\nu W_\mu^- -$$
$$W_\nu^- \partial_\nu W_\mu^+)] - igs_w[\partial_\nu A_\mu(W_\mu^+ W_\nu^- - W_\nu^+ W_\mu^-) - A_\nu(W_\mu^+ \partial_\nu W_\mu^- -$$
$$W_\mu^- \partial_\nu W_\mu^+) + A_\mu(W_\nu^+ \partial_\nu W_\mu^- - W_\nu^- \partial_\nu W_\mu^+)] - \tfrac{1}{2}g^2 W_\mu^+ W_\mu^- W_\nu^+ W_\nu^- +$$
$$\tfrac{1}{2}g^2 W_\mu^+ W_\nu^- W_\mu^+ W_\nu^- + g^2 c_w^2(Z_\mu^0 W_\mu^+ Z_\nu^0 W_\nu^- - Z_\mu^0 Z_\mu^0 W_\nu^+ W_\nu^-) +$$
$$g^2 s_w^2(A_\mu W_\mu^+ A_\nu W_\nu^- - A_\mu A_\mu W_\nu^+ W_\nu^-) + g^2 s_w c_w[A_\mu Z_\nu^0(W_\mu^+ W_\nu^- -$$
$$W_\nu^+ W_\mu^-) - 2A_\mu Z_\mu^0 W_\nu^+ W_\nu^-] - g\alpha[H^3 + H\phi^0\phi^0 + 2H\phi^+\phi^-] -$$
$$\tfrac{1}{8}g^2 \alpha_h[H^4+(\phi^0)^4+4(\phi^+\phi^-)^2+4(\phi^0)^2\phi^+\phi^-+4H^2\phi^+\phi^-+2(\phi^0)^2 H^2] -$$
$$gMW_\mu^+ W_\mu^- H - \tfrac{1}{2}g\tfrac{M}{c_w^2}Z_\mu^0 Z_\mu^0 H - \tfrac{1}{2}ig[W_\mu^+(\phi^0\partial_\mu\phi^- - \phi^-\partial_\mu\phi^0) -$$
$$W_\mu^-(\phi^0\partial_\mu\phi^+ - \phi^+\partial_\mu\phi^0)] + \tfrac{1}{2}g[W_\mu^+(H\partial_\mu\phi^- - \phi^-\partial_\mu H) - W_\mu^-(H\partial_\mu\phi^+ -$$
$$\phi^+\partial_\mu H)] + \tfrac{1}{2}g\tfrac{1}{c_w}(Z_\mu^0(H\partial_\mu\phi^0 - \phi^0\partial_\mu H) - ig\tfrac{s_w^2}{c_w}MZ_\mu^0(W_\mu^+\phi^- - W_\mu^-\phi^+) +$$
$$igs_w MA_\mu(W_\mu^+\phi^- - W_\mu^-\phi^+) - ig\tfrac{1-2c_w^2}{2c_w}Z_\mu^0(\phi^+\partial_\mu\phi^- - \phi^-\partial_\mu\phi^+) +$$
$$igs_w A_\mu(\phi^+\partial_\mu\phi^- - \phi^-\partial_\mu\phi^+) - \tfrac{1}{4}g^2 W_\mu^+ W_\mu^-[H^2 + (\phi^0)^2 + 2\phi^+\phi^-] -$$
$$\tfrac{1}{4}g^2 \tfrac{1}{c_w^2}Z_\mu^0 Z_\mu^0[H^2+(\phi^0)^2+2(2s_w^2-1)^2\phi^+\phi^-] - \tfrac{1}{2}g^2\tfrac{s_w^2}{c_w}Z_\mu^0\phi^0(W_\mu^+\phi^- +$$
$$W_\mu^-\phi^+) - \tfrac{1}{2}ig^2\tfrac{s_w^2}{c_w}Z_\mu^0 H(W_\mu^+\phi^- - W_\mu^-\phi^+) + \tfrac{1}{2}g^2 s_w A_\mu\phi^0(W_\mu^+\phi^- +$$
$$W_\mu^-\phi^+) + \tfrac{1}{2}ig^2 s_w A_\mu H(W_\mu^+\phi^- - W_\mu^-\phi^+) - g^2\tfrac{s_w}{c_w}(2c_w^2-1)Z_\mu^0 A_\mu\phi^+\phi^- -$$
$$g^1 s_w^2 A_\mu A_\mu \phi^+\phi^- - \bar{e}^\lambda(\gamma\partial + m_e^\lambda)e^\lambda - \bar{\nu}^\lambda\gamma\partial\nu^\lambda - \bar{u}_j^\lambda(\gamma\partial + m_u^\lambda)u_j^\lambda -$$
$$\bar{d}_j^\lambda(\gamma\partial + m_d^\lambda)d_j^\lambda + igs_w A_\mu[-(\bar{e}^\lambda\gamma^\mu e^\lambda) + \tfrac{2}{3}(\bar{u}_j^\lambda\gamma^\mu u_j^\lambda) - \tfrac{1}{3}(\bar{d}_j^\lambda\gamma^\mu d_j^\lambda)] +$$
$$\tfrac{ig}{4c_w}Z_\mu^0[(\bar{\nu}^\lambda\gamma^\mu(1+\gamma^5)\nu^\lambda) + (\bar{e}^\lambda\gamma^\mu(4s_w^2 - 1 - \gamma^5)e^\lambda) + (\bar{u}_j^\lambda\gamma^\mu(\tfrac{4}{3}s_w^2 -$$
$$1 - \gamma^5)u_j^\lambda) + (\bar{d}_j^\lambda\gamma^\mu(1 - \tfrac{8}{3}s_w^2 - \gamma^5)d_j^\lambda)] + \tfrac{ig}{2\sqrt{2}}W_\mu^+[(\bar{\nu}^\lambda\gamma^\mu(1+\gamma^5)e^\lambda) +$$
$$(\bar{u}_j^\lambda\gamma^\mu(1+\gamma^5)C_{\lambda\kappa}d_j^\kappa)] + \tfrac{ig}{2\sqrt{2}}W_\mu^-[(\bar{e}^\lambda\gamma^\mu(1+\gamma^5)\nu^\lambda) + (\bar{d}_j^\lambda C_{\lambda\kappa}^\dagger\gamma^\mu(1+$$
$$\gamma^5)u_j^\lambda)] + \tfrac{ig}{2\sqrt{2}}\tfrac{m_e^\lambda}{M}[-\phi^+(\bar{\nu}^\lambda(1-\gamma^5)e^\lambda) + \phi^-(\bar{e}^\lambda(1+\gamma^5)\nu^\lambda)] -$$
$$\tfrac{g}{2}\tfrac{m_e^\lambda}{M}[H(\bar{e}^\lambda e^\lambda) + i\phi^0(\bar{e}^\lambda\gamma^5 e^\lambda)] + \tfrac{ig}{2M\sqrt{2}}\phi^+[-m_d^\kappa(\bar{u}_j^\lambda C_{\lambda\kappa}(1-\gamma^5)d_j^\kappa) +$$
$$m_u^\lambda(\bar{u}_j^\lambda C_{\lambda\kappa}(1+\gamma^5)d_j^\kappa] + \tfrac{ig}{2M\sqrt{2}}\phi^-[m_d^\lambda(\bar{d}_j^\lambda C_{\lambda\kappa}^\dagger(1+\gamma^5)u_j^\kappa) - m_u^\kappa(\bar{d}_j^\lambda C_{\lambda\kappa}^\dagger(1-$$
$$\gamma^5)u_j^\kappa] - \tfrac{g}{2}\tfrac{m_u^\lambda}{M}H(\bar{u}_j^\lambda u_j^\lambda) - \tfrac{g}{2}\tfrac{m_d^\lambda}{M}H(\bar{d}_j^\lambda d_j^\lambda) + \tfrac{ig}{2}\tfrac{m_u^\lambda}{M}\phi^0(\bar{u}_j^\lambda\gamma^5 u_j^\lambda) -$$
$$\tfrac{ig}{2}\tfrac{m_d^\lambda}{M}\phi^0(\bar{d}_j^\lambda\gamma^5 d_j^\lambda) + \bar{X}^+(\partial^2 - M^2)X^+ + \bar{X}^-(\partial^2-M^2)X^- + \bar{X}^0(\partial^2 -$$
$$\tfrac{M^2}{c_w^2})X^0 + \bar{Y}\partial^2 Y + igc_w W_\mu^+(\partial_\mu\bar{X}^0 X^- - \partial_\mu\bar{X}^+ X^0) + igs_w W_\mu^+(\partial_\mu\bar{Y}X^- -$$
$$\partial_\mu\bar{X}^+ Y) + igc_w W_\mu^-(\partial_\mu\bar{X}^- X^0 - \partial_\mu\bar{X}^0 X^+) + igs_w W_\mu^-(\partial_\mu\bar{X}^-Y -$$
$$\partial_\mu\bar{Y}X^+) + igc_w Z_\mu^0(\partial_\mu\bar{X}^+ X^+ - \partial_\mu\bar{X}^- X^-) + igs_w A_\mu(\partial_\mu\bar{X}^+ X^+ -$$
$$\partial_\mu\bar{X}^- X^-) - \tfrac{1}{2}gM[\bar{X}^+ X^+ H + \bar{X}^- X^- H + \tfrac{1}{c_w^2}\bar{X}^0 X^0 H] +$$
$$\tfrac{1-2c_w^2}{2c_w}igM[\bar{X}^+ X^0\phi^+ - \bar{X}^- X^0\phi^-] + \tfrac{1}{2c_w}igM[\bar{X}^0 X^-\phi^+ - \bar{X}^0 X^+\phi^-] +$$
$$igMs_w[\bar{X}^0 X^-\phi^+ - \bar{X}^0 X^+\phi^-] + \tfrac{1}{2}igM[\bar{X}^+ X^+\phi^0 - \bar{X}^- X^-\phi^0]$$

Image Source: Thomas D. Gutierrez

What does it say?

The Standard Model of particle physics is often visualized as a table, similar to the periodic table of elements, and used to describe particle properties, such as mass, charge and spin. It is the theory describing three of the four known fundamental forces (the electromagnetic, weak, and strong interactions, and not including the gravitational force) in the universe, as well as classifying all known elementary particles.

What did it teach us?

Part1

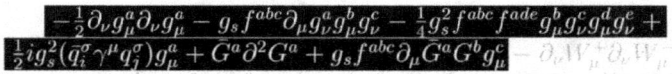

These three lines in the Standard Model are ultraspecific to the gluon, the boson that carries the strong force.

Part2

Almost half of this equation is dedicated to explaining interactions between bosons, particularly W and Z bosons.
Bosons are force-carrying particles, and there are four species of bosons that interact with other particles using three fundamental forces. Photons carry electromagnetism, gluons carry the strong force and W and Z bosons carry the weak force. The most recently discovered boson, the Higgs boson, is a bit different; its interactions appear in the next part of the equation.

$$\tfrac{1}{2}ig_s^2(\bar{q}_i^\lambda \gamma^\mu q_j^\sigma)g_s^a + \bar{G}^a\partial^2 G^a + g_s f^{abc}\partial_\mu \bar{G}^a G^b g_\mu^c - \partial_\nu W_\mu^+ \partial_\nu W_\mu^- -$$
$$M^2 W_\mu^+ W_\mu^- - \tfrac{1}{2}\partial_\nu Z_\mu^0 \partial_\nu Z_\mu^0 - \tfrac{1}{2c_w^2}M^2 Z_\mu^0 Z_\mu^0 - \tfrac{1}{2}\partial_\mu A_\nu \partial_\mu A_\nu - \tfrac{1}{2}\partial_\mu H \partial_\mu H -$$
$$\tfrac{1}{2}m_h^2 H^2 - \partial_\mu \phi^+ \partial_\mu \phi^- - M^2 \phi^+ \phi^- - \tfrac{1}{2}\partial_\mu \phi^0 \partial_\mu \phi^0 - \tfrac{1}{2c_w^2}M\phi^0\phi^0 - \beta_h[\tfrac{2M^2}{g^2} +$$
$$\tfrac{2M}{g}H + \tfrac{1}{2}(H^2 + \phi^0\phi^0 + 2\phi^+\phi^-)] + \tfrac{2M^4}{g^2}\alpha_h - igc_w[\partial_\nu Z_\mu^0(W_\mu^+ W_\nu^- -$$
$$W_\nu^+ W_\mu^-) - Z_\nu^0(W_\mu^+ \partial_\nu W_\mu^- - W_\mu^- \partial_\nu W_\mu^+) + Z_\mu^0(W_\nu^+ \partial_\nu W_\mu^- -$$
$$W_\nu^- \partial_\nu W_\mu^+)] - igs_w[\partial_\nu A_\mu(W_\mu^+ W_\nu^- - W_\nu^+ W_\mu^-) - A_\nu(W_\mu^+ \partial_\nu W_\mu^- -$$
$$W_\mu^- \partial_\nu W_\mu^+) + A_\mu(W_\nu^+ \partial_\nu W_\mu^- - W_\nu^- \partial_\nu W_\mu^+)] - \tfrac{1}{2}g^2 W_\mu^+ W_\mu^- W_\nu^+ W_\nu^- +$$
$$\tfrac{1}{2}g^2 W_\mu^+ W_\nu^- W_\mu^+ W_\nu^- + g^2 c_w^2(Z_\mu^0 W_\mu^+ Z_\nu^0 W_\nu^- - Z_\mu^0 Z_\mu^0 W_\nu^+ W_\nu^-) +$$
$$g^2 s_w^2(A_\mu W_\mu^+ A_\nu W_\nu^- - A_\mu A_\mu W_\nu^+ W_\nu^-) + g^2 s_w c_w[A_\mu Z_\nu^0(W_\mu^+ W_\nu^- -$$
$$W_\nu^+ W_\mu^-) - 2A_\mu Z_\mu^0 W_\nu^+ W_\nu^-] - g\alpha[H^3 + H\phi^0\phi^0 + 2H\phi^+\phi^-] -$$
$$\tfrac{1}{8}g^2 \alpha_h[H^4 + (\phi^0)^4 + 4(\phi^+\phi^-)^2 + 4(\phi^0)^2\phi^+\phi^- + 4H^2\phi^+\phi^- + 2(\phi^0)^2 H^2] -$$
$$gMW_\mu^+ W_\mu^- H - \tfrac{1}{2}g\tfrac{M}{c_w^2}Z_\mu^0 Z_\mu^0 H - \tfrac{1}{2}ig[W_\mu^+(\phi^0\partial_\mu\phi^- - \phi^-\partial_\mu\phi^0) -$$
$$W_\mu^-(\phi^0\partial_\mu\phi^+ - \phi^+\partial_\mu\phi^0)] + \tfrac{1}{2}g[W_\mu^+(H\partial_\mu\phi^- - \phi^-\partial_\mu H) - W_\mu^-(H\partial_\mu\phi^+ -$$
$$\phi^+\partial_\mu H)] + \tfrac{1}{2}g\tfrac{1}{c_w}(Z_\mu^0(H\partial_\mu\phi^0 - \phi^0\partial_\mu H) - ig\tfrac{s_w^2}{c_w}MZ_\mu^0(W_\mu^+\phi^- - W_\mu^-\phi^+) +$$
$$igs_w MA_\mu(W_\mu^+\phi^- - W_\mu^-\phi^+) - ig\tfrac{1-2c_w^2}{2c_w}Z_\mu^0(\phi^+\partial_\mu\phi^- - \phi^-\partial_\mu\phi^+) +$$
$$igs_w A_\mu(\phi^+\partial_\mu\phi^- - \phi^-\partial_\mu\phi^+) - \tfrac{1}{4}g^2 W_\mu^+ W_\mu^-[H^2 + (\phi^0)^2 + 2\phi^+\phi^-] -$$
$$\tfrac{1}{4}g^2\tfrac{1}{c_w^2}Z_\mu^0 Z_\mu^0[H^2 + (\phi^0)^2 + 2(2s_w^2 - 1)^2\phi^+\phi^-] - \tfrac{1}{2}g^2\tfrac{s_w^2}{c_w}Z_\mu^0\phi^0(W_\mu^+\phi^- +$$
$$W_\mu^-\phi^+) - \tfrac{1}{2}ig^2\tfrac{s_w^2}{c_w}Z_\mu^0 H(W_\mu^+\phi^- - W_\mu^-\phi^+) + \tfrac{1}{2}g^2 s_w A_\mu\phi^0(W_\mu^+\phi^- +$$
$$W_\mu^-\phi^+) + \tfrac{1}{2}ig^2 s_w A_\mu H(W_\mu^+\phi^- - W_\mu^-\phi^+) - g^2\tfrac{s_w}{c_w}(2c_w^2 - 1)Z_\mu^0 A_\mu\phi^+\phi^- -$$
$$g^1 s_w^2 A_\mu A_\mu \phi^+\phi^-$$

Part3

This part of the equation describes how elementary matter particles interact with the weak force. According to this formulation, matter particles come in three generations, each with different masses. The weak force helps massive matter particles decay into less massive matter particles.

This section also includes basic interactions with the Higgs field, from which some elementary particles receive their mass.

$$-\bar{e}^\lambda(\gamma\partial + m_e^\lambda)e^\lambda - \bar{\nu}^\lambda\gamma\partial\nu^\lambda - \bar{u}_j^\lambda(\gamma\partial + m_u^\lambda)u_j^\lambda - \bar{d}_j^\lambda(\gamma\partial + m_d^\lambda)d_j^\lambda + igs_w A_\mu[-(\bar{e}^\lambda\gamma^\mu e^\lambda) + \tfrac{2}{3}(\bar{u}_j^\lambda\gamma^\mu u_j^\lambda) - \tfrac{1}{3}(\bar{d}_j^\lambda\gamma^\mu d_j^\lambda)] + \tfrac{ig}{4c_w}Z_\mu^0[(\bar{\nu}^\lambda\gamma^\mu(1+\gamma^5)\nu^\lambda) + (\bar{e}^\lambda\gamma^\mu(4s_w^2 - 1 - \gamma^5)e^\lambda) + (\bar{u}_j^\lambda\gamma^\mu(\tfrac{4}{3}s_w^2 - 1 - \gamma^5)u_j^\lambda) + (\bar{d}_j^\lambda\gamma^\mu(1 - \tfrac{8}{3}s_w^2 - \gamma^5)d_j^\lambda)] + \tfrac{ig}{2\sqrt{2}}W_\mu^+[(\bar{\nu}^\lambda\gamma^\mu(1+\gamma^5)e^\lambda) + (\bar{u}_j^\lambda\gamma^\mu(1+\gamma^5)C_{\lambda\kappa}d_j^\kappa)] + \tfrac{ig}{2\sqrt{2}}W_\mu^-[(\bar{e}^\lambda\gamma^\mu(1+\gamma^5)\nu^\lambda) + (\bar{d}_j^\kappa C_{\lambda\kappa}^\dagger\gamma^\mu(1+\gamma^5)u_j^\lambda)] + \tfrac{ig}{2\sqrt{2}}\tfrac{m_e^\lambda}{M}[-\phi^+(\bar{\nu}^\lambda(1-\gamma^5)e^\lambda) + \phi^-(\bar{e}^\lambda(1+\gamma^5)\nu^\lambda)] -$$

Part4

$$\gamma^5)u_j^\lambda)] + \tfrac{ig}{2\sqrt{2}}\tfrac{m_e^\lambda}{M}[-\phi^+(\bar{\nu}^\lambda(1-\gamma^5)e^\lambda) + \phi^-(\bar{e}^\lambda(1+\gamma^5)\nu^\lambda)] - \tfrac{g}{2}\tfrac{m_e^\lambda}{M}[H(\bar{e}^\lambda e^\lambda) + i\phi^0(\bar{e}^\lambda\gamma^5 e^\lambda)] + \tfrac{ig}{2M\sqrt{2}}\phi^+[-m_d^\lambda(\bar{u}_j^\lambda C_{\lambda\kappa}(1-\gamma^5)d_j^\kappa) + m_u^\lambda(\bar{u}_j^\lambda C_{\lambda\kappa}(1+\gamma^5)d_j^\kappa)] + \tfrac{ig}{2M\sqrt{2}}\phi^-[m_d^\lambda(\bar{d}_j^\lambda C_{\lambda\kappa}^\dagger(1+\gamma^5)u_j^\kappa) - m_u^\lambda(\bar{d}_j^\lambda C_{\lambda\kappa}^\dagger(1-\gamma^5)u_j^\kappa)] - \tfrac{g}{2}\tfrac{m_u^\lambda}{M}H(\bar{u}_j^\lambda u_j^\lambda) - \tfrac{g}{2}\tfrac{m_d^\lambda}{M}H(\bar{d}_j^\lambda d_j^\lambda) + \tfrac{ig}{2}\tfrac{m_u^\lambda}{M}\phi^0(\bar{u}_j^\lambda\gamma^5 u_j^\lambda) - \tfrac{ig}{2}\tfrac{m_d^\lambda}{M}\phi^0(\bar{d}_j^\lambda\gamma^5 d_j^\lambda) + \bar{X}^+(\partial^2 - M^2)X^+ + \bar{X}^-(\partial^2 - M^2)X^- + \bar{X}^0(\partial^2 - $$

This part of the equation describes how matter particles interact with Higgs ghosts, virtual artifacts from the Higgs field.

Part5

$$\frac{ig}{\lambda}\frac{m_h^2}{M}\phi^0(d_j^\lambda \gamma^5 d_j^\lambda) + \bar{X}^+(\partial^2 - M^2)X^+ + \bar{X}^-(\partial^2 - M^2)X^- + \bar{X}^0(\partial^2 - \frac{M^2}{c_w^2})X^0 + \bar{Y}\partial^2 Y + igc_w W_\mu^+(\partial_\mu \bar{X}^0 X^- - \partial_\mu \bar{X}^+ X^0) + igs_w W_\mu^+(\partial_\mu \bar{Y} X^- - \partial_\mu \bar{X}^+ Y) + igc_w W_\mu^-(\partial_\mu \bar{X}^- X^0 - \partial_\mu \bar{X}^0 X^+) + igs_w W_\mu^-(\partial_\mu \bar{X}^- Y - \partial_\mu \bar{Y} X^+) + igc_w Z_\mu^0(\partial_\mu \bar{X}^+ X^+ - \partial_\mu \bar{X}^- X^-) + igs_w A_\mu(\partial_\mu \bar{X}^+ X^+ - \partial_\mu \bar{X}^- X^-) - \frac{1}{2}gM[\bar{X}^+ X^+ H + \bar{X}^- X^- H + \frac{1}{c_w^2}\bar{X}^0 X^0 H] + \frac{1-2c_w^2}{2c_w}igM[\bar{X}^+ X^0 \phi^+ - \bar{X}^- X^0 \phi^-] + \frac{1}{2c_w}igM[\bar{X}^0 X^- \phi^+ - \bar{X}^0 X^+ \phi^-] + igMs_w[\bar{X}^0 X^- \phi^+ - \bar{X}^0 X^+ \phi^-] + \frac{1}{2}igM[\bar{X}^+ X^+ \phi^0 - \bar{X}^- X^- \phi^0]$$

This last part of the equation includes more ghosts. These ones are called Faddeev-Popov ghosts, and they cancel out redundancies that occur in interactions through the weak force.

But was it practical?

The Standard Model predicted the existence of the W and Z bosons, the gluon, the top and charm quarks and the Higgs boson. The predictions were experimentally confirmed with good precision.

42 The first Friedmann equation

$$H^2 = \left(\frac{\dot{a}}{a}\right)^2 = \frac{8\pi G}{3}\rho - \frac{kc^2}{a^2} + \frac{\Lambda c^2}{3}$$

What does it say?

The universe is not static, but rather expands or contracts depending on what the expansion rate and contents of the universe are.

What did it teach us?

According to Ethan siegel, " the Friedmann equation describes how, based on what is in the universe, its expansion rate will change overtime. If you want to know where the universe came from and where it's headed, all you need to measure is how it is expanding today and what is in it. This equation allows you to predict the rest"

But was it practical?

It lead to the Big Bang theory, it tells you how the fabric of the universe expands or contracts as a function of time, and the development of the Hubble telescope

History:

It was first derived by Alexander Friedmann in 1922 from Einstein's General relativity.

43 The Callan-Symanzik equation

$$\left[M\frac{\partial}{\partial M} + \beta(g)\frac{\partial}{\partial g} + n\gamma\right] G^n(x_1, x_2, \ldots, x_n; M, g) = 0$$

What does it say?

According to Matt Strassler it is a vital first-principles equation from 1970, essential for describing how naïve expectations will fail in a quantum world.

What did it teach us?

Because tiny quantum fluctuations slightly alter a strong nuclear force's dependence on distance, the strong force is prevented from decreasing at long distances. What the Callan-Symanzik equation does is relate this dramatic and difficult to calculate effect, important when the distance is roughly the size of a proton, to more subtle but easier to calculate effects that can be measured when the distance is much smaller than a proton.

But was it practical?

The equation has numerous applications, including allowing physicists to estimate the mass and size of the proton and neutron, which make up the nuclei of atoms. It was also used to understand asymptotic freedom.

History:

The equation was discovered independently by Curtis Callan and Kurt Symanzik in 1970.

44 The minimal surface equation

$$\mathcal{A}(u) = \int (1 + |\nabla_u|^2)^{1/2} dx_1 \ldots dx_r$$

What does it say?

The minimal surface equation somehow encodes the beautiful soap films that form on wire boundaries when you dip them in soapy water. It is applied in molecular engineering, material science, endoplasmic reticulum, general relativity, architecture and art world.

45 The Heat equation

$$\frac{\partial u}{\partial t} = \alpha \left(\frac{\partial^2 u}{\partial x^2} + \frac{\partial^2 u}{\partial y^2} + \frac{\partial^2 u}{\partial z^2} \right)$$

What does it say?

It describes how the distribution of heat evolves over time in a solid medium, as it spontaneously flows from places where it is higher towards places where it is lower.

What did it teach us?

In financial mathematics it is used to solve the Black-Scholes partial differential equation. In quantum mechanics for finding spread of wavefunction in potential free region, it was also instrumental in the solution of longtanding Poincare conjecture of topology.

But was it practical?

The equation has applications in financial mathematics, image analysis and in machine learning it is the driving theory behind scale space or graph Laplacian methods.

History:

The equation was first developed and solved by Joseph Fourier in 1822 to describe heat flow.

56 The Gaussian Integral

$$\int_{-\infty}^{\infty} e^{-x^2} dx = \sqrt{\pi}$$

The function e^{-x^2} in itself is a very ugly function to integrate, but when done across the entire real line, i.e. from minus infinity to infinity, it gives a bizarrely clean answer. It is certainly not obvious at first glance that the area under the curve is the square root of pi.

This formula is of extreme importance in statistics, as it represents the normal distribution.

47 The Basel Identity

$$1 + \frac{1}{4} + \frac{1}{9} + \frac{1}{16} + \frac{1}{25} + \cdots = \frac{\pi^2}{6}$$

What does it say?

It is an infinite series that produces pi.

What did it teach us?

This equation says that if you take the reciprocal of all the square numbers, and then add them all together, you get pi squared over six. This was proved by Euler.

But was it practical?

It is a great example of unexpected patterns that appear in mathematics and reveal a deeper logic linking an infinite series to the circle constant.

History:

It was first proposed by Pietro Mengoli in 1650 and solved by Leonhard Euler in 1734 and read on the 5th December 1735 in the saint Petersburg Academy of sciences.

48 The KFC formula

DETAILS OF THE MIXTURE

One chicken, cut up, cut the breast pieces in half for more even frying.

Bowl Number one.
a) 2 cups of purpose flour
b) 2/3 table spoon salt
c) ½ tablespoon dried thyme leaves
d) ½ tablespoon dried basil leaves
e) 1/3 tablespoon dried oregano leaves
f) 1 tablespoon celery salt
g) 1 tablespoon ground black pepper
h) 1 tablespoon dried mustard
i) 4 tablespoons paprika
j) 2 tablespoons garlic salt
k) 1 tablespoons ground white pepper

Bowl Number two.
a) 1 cup butter milk
b) 1 egg beaten

RECIPE

1. Cut the chicken up into frying pieces

2. Combine beaten egg 1-2 per chicken with butter milk 1-2 cups placed in a large bowl, soak entirely the chicken pieces inside.

3. In a separate bowl, mix flour 2 cups with the herb-spice (black pepper, salt, thyme, basil…..). Use a scoop to remove the chicken from the butter -milk egg bath into the flour mixture. Use hands to mix.

4. Heat about 3inches of the oil in a heavy pot over a pressure cooker, heat to 350degrees. (use deep frying thermometer to check the temperature). Fry until medium golden brown chicken .

5. Sprinkle a little MSG on the finished chicken pieces before eating.

dropped in the **pressure cooker**. You have to cook it until it **turns brown**. Then you put the lid on the pressure cooker and bring it to 12 pounds of pressure for 10 minutes. And then you start letting the pressure off, and when you have uncapped it and the pressure is off, it will be perfect: golden brown and falling-off-the-bone.

History

The 11 herbs and spices that make up the secret recipe used by KFC where found by Harland Sanders in Kentucky in the late 1930s. In 1939 Sanders found that using a pressure cooker produced tasty, moist chicken in eight or nine minutes better than pan frying and deep frying which takes 30minutes per order. By July 1940 Sanders finalized what came to be known as his original recipe. Finally the company was established in 1950.

49 The Coca-Cola formula

RECIPE

1. Fluid extract of coca: 3 drams USP
2. Citric acid: 3 oz
3. Caffeine: 1 oz
4. Sugar: 30 lb
5. Water: 2.5 gallons
6. Lime juice: 2 pints, 1 quart
7. Vanilla: 1 oz
8. Caramel: 1.5 oz or more to colour

Into every five gallons of syrup, add 2oz of the following seven-part flavouring:

9. Alcohol: 8 oz
10. Orange oil: 20 drops
11. Lemon oil: 30 drops
12. Nutmeg oil: 10 drops
13. Coriander oil : 5 drops
14. Neroli oil: 10 drops
15. Cinnamon oil: 10 drops

History

Coca-Cola was invented by John Pemberton who shared the formula to at least four people before his death in 1888. The formula was later bought by Asa Candler in 1891 from Pemberton's estate and formed it into the Coca-Cola Company.

50 The Kalman Filter

$$\hat{X}_k = K_k \cdot Z_k + (1 - K_k) \cdot \hat{X}_{k-1}$$

$\hat{X}_k$-current estimation
K_k-Kalman gain
Z_k-measurement value
$\hat{X}_{k-1}$-Kalman gain

What does it say?

The Kalman filter is an algorithm which estimates a process by using a form of feedback control. It is a recursive data processing algorithm.

What did it teach us?

To estimate the state of a system from measurements which contain random errors. An example is estimating the position and velocity of a satellite from radar data.

But was it practical?

A common application of the filter is for guidance, navigation, and control of vehicles, particularly aircraft, spacecraft and dynamically positioned ships. Other applications include: Autopilot, brain computer interface, economics, nuclear medicine, radar tracker, weather forecasting, 3D modeling, structural health monitoring.

History:

The filter is named after Rudolph E.Kalman, who in 1960 published his famous paper describing a recursive solution to the discrete data linear filtering problem.

My Favorite Formula

$$\rho = \frac{F^2}{8\pi\alpha\hbar c}$$

I developed this formula some years ago. It is a summarized equation for the standard model of physics. The equation represents the coupling constant (α) as a function of the energy density (ρ) for any force (F) exerted in an interaction. It is the idea behind the Franzl Aus Tirol curve/ sketch which he submitted to Wikipedia on 29 July 2012 and featured in the "Coupling constant".

When the equation is combined with the area law of Einstein General relativity (Friedmann Equation $A = \frac{c^4}{2G\rho}$) that is, when we substitute for ρ, we get the Area law that captures the quantum discreteness of geometry as, $A = \frac{4\pi\alpha\hbar c^5}{GF^2}$, such that when the force exerted in an interaction is the gravitational force $F = \frac{c^4}{G}$, the geometry of space is nologer continuous but granular in the form, $A = \frac{4\pi\alpha\hbar G}{c^3} \approx 4\pi\alpha l_p^2$. This final block is the founding block of Quantum gravity and has been derived in Loop quantum gravity.

Bibliography/References

Sen, D.(2014). "The Uncertainty relations in quantum mechanics". Current Science. 107 (2): 203-218

See Brian Pendleton: Quantum theory 2012/2013, section 4.3 The Dirac Equation

Jacques H.H. Perk and Helen Au-Yang, (2006)"Yang-Baxter Equations", arXiv:math-ph/0606053

Huerta, John (2010) "Introducing the Quaternions"PDF

Roger Temam (1984): "Navier-Stokes Equations: Theory and Numerical Analysis", ACM Chelsea Publishing, ISBN 978-0-8218-2737-6

Kendig, Keith (2000). "Is a 2000-Year-Old Formula Still Keeping some screts?" Amer. Math. Monthly. 107:402-415. Doi:10.2307/2695295

Weisstein, Eric W. "Heron's Formula" MathWorld

Weinberg, S. (2013). Lectures on Quantum Mechanics, Cambridge UK, ISBN 978-1-107-02872-2.

"Bernoulli's Equation" NASA Glenn Research Center. Retrieved 2009-03-04.

Boeing (2015) "Chaos Theory and the Logistic Map" Journal of the Optical Society of America B. 3(5):741.

Moran; Shapiro (2000). Fundamentals of Engineering Thermodynamics (4th ed) Wiley. ISBN0-471.31713-6.

A. Sachs, A. Goetze, and O. Neugebauer. *Mathematical Cuneiform Texts*, American Oriental Society, New Haven 1945.

Massimo Franceschet(2010). "PageRank:Standing on the shoulders of giants". arXiv: 1002.2858

"EgdeRank: The Secret Sauce That Makes Facebook's News Feed Tick" techcrunch.com (2010)

Ian Stewart (2012) "Seventeen Equations that changed the World" https://pdfs.semanticscholar.org PDF

Guillen Michael (1995) "Five Equations that Changed the World" http://web.mit.edu PDF

Note: (Equation 41) Thomas Gutierrez, an assistant professor of Physics at California Polytechnic State University, transcribed the Standard Model Lagrangian for the web. He derived it from Diagrammatica, a theoretical physics reference written by Nobel Laureate Martinus Veltman. In Gutierrez's dissemination of the transcript, he noted a sign error he made somewhere in the equation. Good luck finding it!

Balungi Francis (2018) "Quantum Gravity in a Nutshell1" Book

www.ingramcontent.com/pod-product-compliance
Lightning Source LLC
Chambersburg PA
CBHW070435220526
45466CB00004B/1689